Raccoon Dogs

Ruby Daniels

Big Buddy Books

An Imprint of Abdo Publishing
abdobooks.com

abdobooks.com

Published by Abdo Publishing, a division of ABDO, PO Box 398166, Minneapolis, Minnesota 55439. Copyright © 2025 by Abdo Consulting Group, Inc. International copyrights reserved in all countries. No part of this book may be reproduced in any form without written permission from the publisher. Big Buddy Books™ is a trademark and logo of Abdo Publishing.

Printed in the United States of America, North Mankato, Minnesota
102024
012025

THIS BOOK CONTAINS RECYCLED MATERIALS

Design: Elena Klinkner, Mighty Media, Inc.
Production: Mighty Media, Inc.
Editor: Liz Salzmann
Cover Photograph: nojafoto/Shutterstock Images
Interior Photographs: Alonephotoshoot/Shutterstock Images, p. 29; Christian Musat/Adobe Stock, p. 7; Jurgen_Schmidt/Shutterstock Images, p. 5; Kacper Pawlik/Shutterstock Images, p. 11; Kazu/Adobe Stock, p. 14 (bottom); kuremo/Adobe Stock, p. 21; Mario Plechaty/Adobe Stock, p. 14 (top); Mikhail/Adobe Stock, p. 15 (bottom); Mircea Costina/Adobe Stock, p. 15 (middle); photoPepp/Adobe Stock, p. 13; russell102/Adobe Stock, p. 23; Stanislav Duben/Adobe Stock, p. 17; Stanislav Duben/Shutterstock Images, p. 25; stas111/Adobe Stock, p. 18 (compass rose); tanarch/Adobe Stock, pp. 18–19 (maps); Vaclav Matous/Shutterstock Images, p. 27; Zanna Pesnina/Shutterstock Images, p. 9
Design Elements: flovie/Shutterstock Images (patchwork pattern); Mighty Media, Inc. (series logos & icons)

Library of Congress Control Number: 2024938324

Publisher's Cataloging-in-Publication Data
Names: Daniels, Ruby, author.
Title: Raccoon Dogs / by Ruby Daniels
Description: Minneapolis, Minnesota : ABDO Publishing, 2025 | Series: Odd but adorable animals | Includes online resources and index.
Identifiers: ISBN 9781098295158 (lib. bdg.) | ISBN 9798384915201 (ebook)
Subjects: LCSH: Raccoon dog--Juvenile literature. | Carnivores--Juvenile literature. | Foxes--Juvenile literature. | Mammals--Juvenile literature. | Invasive organisms—Juvenile literature. | Curiosities and wonders--Juvenile literature.
Classification: DDC 599.772--dc23

Contents

A Raccoon Dog Encounter............ 4
A Closer Look........................ 6
Life as a Raccoon Dog 8
Curious Canids......................12
Creature Feature 14
Happy Homes 16
Location Station 18
Critter Culture 20
Amazing Investigations 24
Threats and Hope................... 26
Odd or Adorable?................... 28
Glossary........................... 30
Online Resources 31
Index 32

A Raccoon Dog Encounter

You're enjoying a peaceful evening hike in the forests of Hokkaido, Japan. Suddenly, something steps onto the path in front of you. It has black fur on its face like a raccoon. But its tail has a black tip rather than rings. You've just spotted an odd but adorable raccoon dog!

The black fur on a raccoon dog's face reduces glare. This allows the animal to see more clearly in the sun.

5

A Closer Look

Raccoon dogs are not related to raccoons. Like dogs, they are part of the **Canidae** family. There are two types of raccoon dogs. Japanese raccoon dogs live in Japan. Common raccoon dogs live in other parts of the world. Raccoon dogs weigh about 16.5 pounds (7.5 kg).

Raccoon dogs are 20 to 27 inches (50 to 69 cm) long.

Life as a Raccoon Dog

Raccoon dogs eat small animals, fish, fruits, and plants. They can be active at any time, though they're most often active at night. Raccoon dogs live in pairs and sometimes in small groups. They poop in the same areas. Scientists believe they use these areas to communicate with other raccoon dogs.

Raccoon dogs often groom each other and sleep near each other.

9

Female raccoon dogs have a litter of pups once a year. The male will bring his **pregnant** mate food. After the litter is born, the pair raises their young together. One watches the pups while the other looks for food. The pups are usually independent after four or five months.

Raccoon dogs have five to seven pups in a litter.

Curious Canids

Raccoon dogs are one of the only **canids** that can climb trees. This is due to their curved claws. Raccoon dogs are also the only canid that enters a **hibernation**-like state called **torpor** in winter. And raccoon dogs don't bark the way many canids do. They growl, **mewl**, and make high-pitched whining calls.

Raccoon dogs growl when they are threatened.

Creature Feature

There are white raccoon dogs. But they are very rare.

The Oklahoma City Zoo is the only **accredited** zoo in the US that houses raccoon dogs.

Raccoon dogs are illegal to own as pets in the US. They can harm people, other animals, and the **environment**.

It's possible that raccoon dogs look like raccoons due to convergent **evolution**. This is when two different **species** evolve in similar ways to adapt to similar **habitats**.

Neoguri is the Korean word for raccoon dog. There is a Korean brand of ramen noodles called *Neoguri*.

Happy Homes

Both **species** of raccoon dog are native to eastern Asia. They often live in forests or thick grasses near water. In their native homes, they help control populations of smaller animals. They also spread seeds from foods they have eaten. In the 1900s, raccoon dogs were brought to Europe for the fur trade. They have become an **invasive** species there.

Raccoon dogs grow long, thick fur for the winter. It helps keep the animals warm.

Location Station

Critter Culture

In Japan, raccoon dogs are called *tanuki*. They are an important part of Japanese **culture**. Many Japanese folktales feature tanuki. They are also said to bring good financial luck. For this reason, there are tanuki statues outside many Japanese shops. There are even **shrines** honoring tanuki in Japan.

Tamon-Ji is a temple near Tokyo, Japan. It is called the Tanuki Temple because of its role in a legend about tanuki.

Enthusiasm for tanuki has spread around the world. Some characters in Nintendo's *Super Mario* games wear Tanooki Suits inspired by tanuki. The characters Tom Nook, Timmy, and Tommy in *Animal Crossing* video games are tanuki. And the 1994 Japanese movie *Pom Poko* is about a group of tanuki trying to save their **habitat**.

Tom Nook is known as Tanukichi in Japan.

Amazing Investigations

Scientists study how raccoon dogs affect **habitats** where they are not native. Some experts believe the virus that causes **COVID-19** started in raccoon dogs. The raccoon dogs were being sold for fur and meat at a market in China. However, other experts disagree. Scientists still don't know where the virus originally came from.

Scientists also study how raccoon dogs adapt to different habitats.

Threats and Hope

Raccoon dogs are not considered **endangered**. But they face **threats**. They are hunted for fur and meat. Raccoon dogs are also frequent victims of road accidents. There is hope for raccoon dogs facing these threats. Hunting them is limited in some places. And many organizations are working to end the fur trade.

A raccoon dog's natural predators include wolves, lynx (*pictured*), eagles, and more.

Odd or Adorable?

Raccoon dogs are unusual and beloved animals. They look strange, but they're also considered cute by many people around the world. What do you think makes an animal odd? What makes an animal adorable? Do you think raccoon dogs are odd, adorable, or both? Why?

Raccoon dogs can live up to 14 years in captivity.

Glossary

accredited—having been given official approval based on set standards.

Canidae (KAN-uh-dee)—the scientific Latin name for the dog family. Members of this family are called canids.

COVID-19—a serious illness that first appeared in late 2019.

culture (KUHL-chuhr)—the arts, beliefs, and ways of life of a group of people.

endangered—having few left in the world.

environment—the natural world, including air, water, land, and animals.

evolve—to change or develop slowly over time. The process of evolving is evolution.

habitat—a place where a living thing is naturally found.

hibernation—when an animal spends the winter sleeping.

invasive—spreading to a non-native habitat and causing harm there.

mewl—to cry weakly.

pregnant—having one or more babies growing in the body.
shrine—a structure built to honor a saint or god.
species (SPEE-sheez)—living things that are very much alike.
threat—something that could be harmful.
torpor—a state of slowed body and mind activity.

Online Resources

To learn more about raccoon dogs, please visit **abdobooklinks.com** or scan this QR code. These links are routinely monitored and updated to provide the most current information available.

Index

Animal Crossing, 22, 23
Asia, 16, 18, 19

canids, 6, 12
China, 18, 19, 24
claws, 12
COVID-19, 24
culture, 20, 21, 22, 23

endangerment, 26
Europe, 16, 18
evolution, 15

folktales, 20, 21
food, 8, 10, 16
forests, 4, 16
fur, 4, 5, 17, 24, 26

fur trade, 16, 24, 26
habitats, 15, 16, 18, 19, 22, 24, 25
Hokkaido, Japan, 4
hunting, 26

Japan, 4, 6, 18, 19, 20, 21, 22, 23

map, 18, 19
mates, 10

Neoguri (ramen), 15

Oklahoma City Zoo, 14

Pom Poko, 22
pups, 10, 11

raccoons, 4, 6, 15
shrines, 20, 21
sounds, 12, 13
statues, 20, 21
studies, 24, 25
Super Mario, 22

tails, 4
threats, 26, 27
torpor, 12

United States, 14, 15, 18

weight, 6
white raccoon dogs, 14